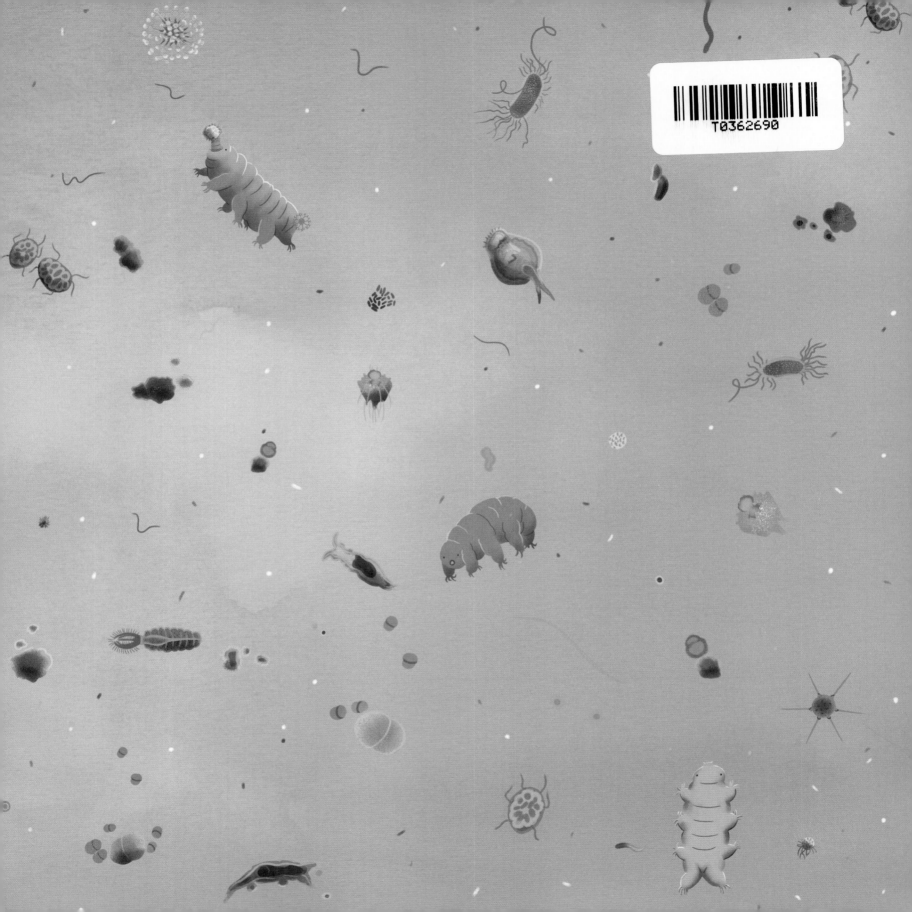

For Logan and Briella – AM

For my husband Luke, and his infectious love of learning – JF

 A catalogue record for this book is available from the National Library of Australia

ISBN: 9781486316052 (hbk)
ISBN: 9781486316069 (epdf)
ISBN: 9781486316076 (epub)

Published by:
CSIRO Publishing
36 Gardiner Road, Clayton VIC 3168
Private Bag 10, Clayton South VIC 3169
Australia

Telephone: +61 3 9545 8400
Email: publishing.sales@csiro.au
Website: www.publish.csiro.au
Sign up to our email alerts: publish.csiro.au/earlyalert

Edited by Nan McNab
Cover, text design and layout by Jennifer Falkner
Printed in China by Toppan Leefung Printing Limited

Tardigrade anatomy illustration based on an image by zombiu26/Adobe Stock.

The views expressed in this publication are those of the author and illustrator and do not necessarily represent those of, and should not be attributed to, the publisher or CSIRO.

CSIRO acknowledges the Traditional Owners of the lands that we live and work on across Australia and pays its respect to Elders past and present. CSIRO recognises that Aboriginal and Torres Strait Islander peoples have made and will continue to make extraordinary contributions to all aspects of Australian life including culture, economy and science. CSIRO is committed to reconciliation and demonstrating respect for Indigenous knowledge and science. The use of Western science in this publication should not be interpreted as diminishing the knowledge of plants, animals and environment from Indigenous ecological knowledge systems.

The author acknowledges the assistance of Dr Sandra Claxton, Dr Thomas Boothby, Courtney Hunt, Melanie Dana Borup, and the Tasmanian Herbarium in developing this manuscript.

Note for readers: Words in bold are explained in the glossary at the back of the book.

Note for teachers: Teacher notes are available at: https://www.publish.csiro.au/book/8071/#forteachers

Tardigrades
NATURE'S TOUGHEST SURVIVORS

WRITTEN BY

Anne Morgan

ILLUSTRATED BY

Jennifer Falkner

CSIRO

PUBLISHING

Planet Earth is home to some very tough animals.

Some of them have survival skills that we humans can only dream about.

But one kind of creature can survive extreme dangers that would kill almost every other **organism**. They must be *World Champion Survivors!*

Don't be fooled by their chubby bodies and stubby legs.

These creatures have been found near the top of one of the world's highest mountains.

Humans need **oxygen** tanks to climb to those heights.

These creatures can live near the bottom of the world's deepest ocean.

It's pitch black down there, and the pressure would crush just about any other living thing.

Humans can only explore those depth in **submersibles**.

If astronauts didn't wear special suits in space, the lack of air, the **cosmic radiation** and the extreme cold would quickly kill them.

Yet some of these Super Survivors once lasted for 10 days while being exposed to all the hazards of space!

Can you guess what these animals are yet?

No, not those beasts!

They are called *tardigrades*. They're so tiny, you'll need magnification to see them clearly.

Scientists use a **microscope** to get a good view of tardigrades, but may need to magnify them up to 100 times.

Tardigrades might look like caterpillars in sleeping bags, but they're not insects.

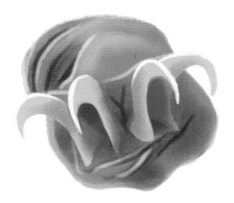

They're often called 'water bears', because they have snouts and claws and a slow, clumsy way of walking in water.

Their other cute nickname is 'moss piglets'. That's because they're often found in the water that clings to moss.

But they're not bears or pigs. They're not fish, either. They are tiny, eight-legged animals that aren't closely related to any other animals.

They might look cute from a distance, but if you were small enough to be their dinner, you would find them terrifying!

Tardigrade **species** may eat **algae**, **bacteria**, tiny **fungi** or other **microorganisms**.

Some will even eat smaller tardigrades!

Tardigrades don't have teeth. Hidden inside their heads are two sharp, sword-shaped **'stylets'**.

When they eat, they push these stylets out of their mouths to spear their food.

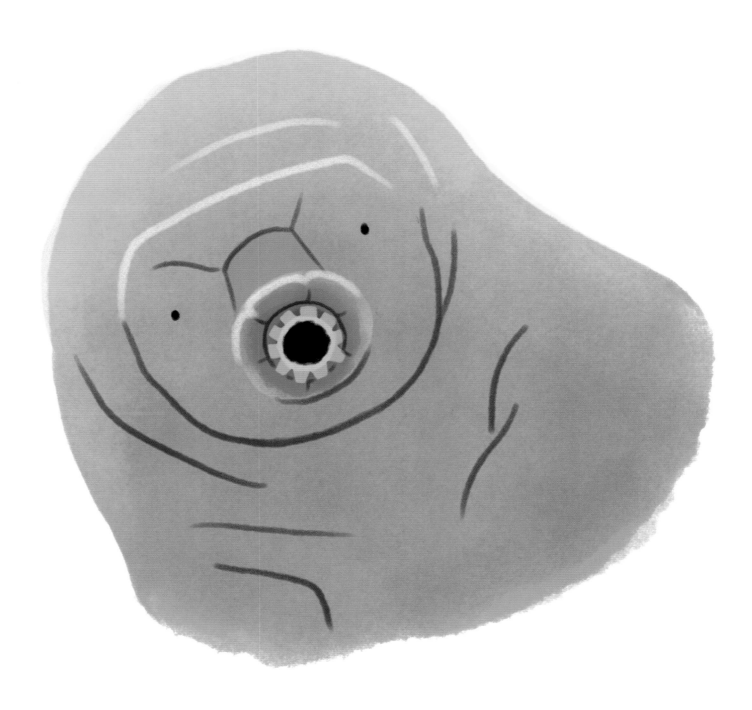

Their round, muscly mouths work like vacuum-cleaner tubes,
slurping juices from the **cells** of other microorganisms.

The ancestors of today's tardigrades appeared on Earth about 600 million years ago.

Dinosaurs didn't appear until about 400 million years later.

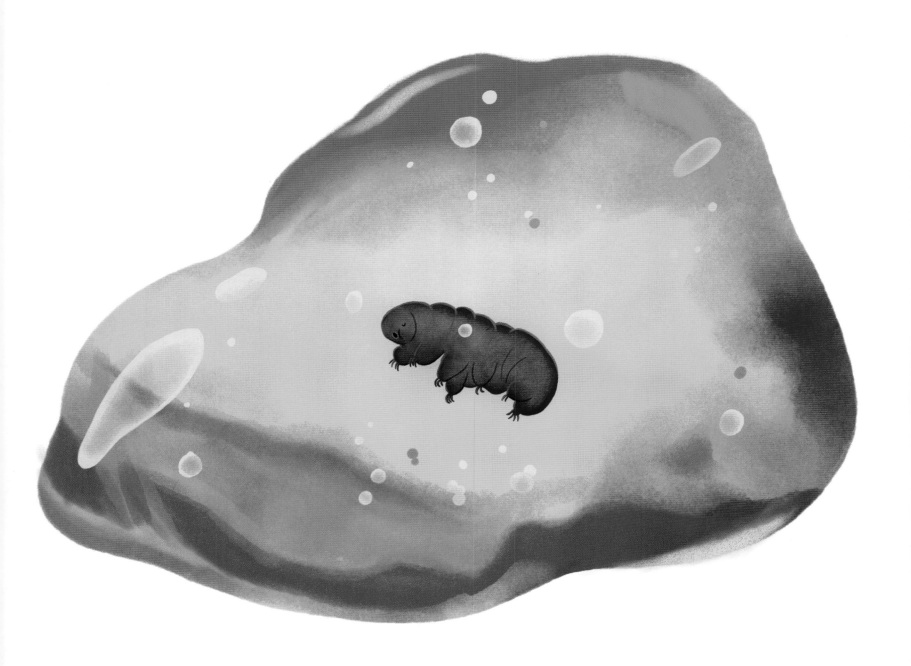

The mighty dinosaurs went extinct, but the teeny, tiny tardigrades survived!

The tardigrades' skin is called their **'cuticle'**. When they grow, they shed their old cuticles, and replace them with bigger ones.

If the water around them becomes too warm, or too cold, for comfort, they grow extra cuticles over their existing ones. Then they turn into tough little **cysts**.

The tardigrades go to sleep inside these thick, protective coats and wait until the temperature becomes comfortable again. Then they crawl out of their tough old shells.

But the tardigrades have an even more astounding trick.

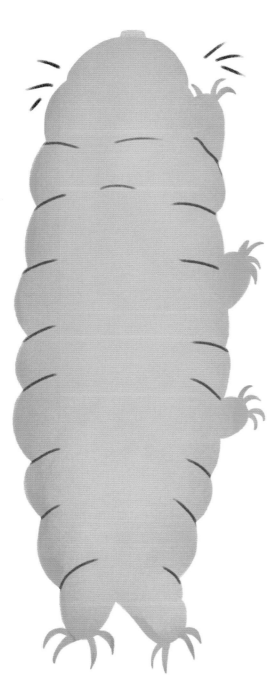

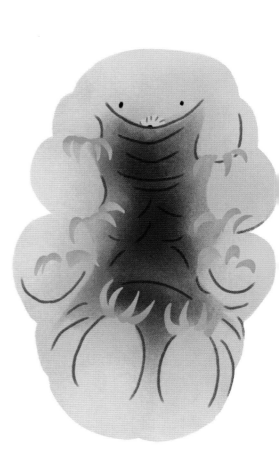

  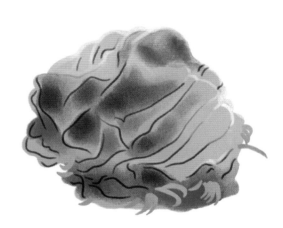

When they are under extreme stress, they roll over, pull in their legs and roll up into balls called 'tuns'. Then they dry out and go into a deep, deep sleep, in which they are nearly, but not quite, dead.

Most tardigrades live for less than a year. But when they become tuns, they can go into **suspended animation** – sometimes for many years.

When they are no longer stressed, the tuns come out of suspended animation and change back into normal tardigrades.

These tough little critters are so light that the wind can blow them, and their eggs, far and wide.

After their eggs hatch, a new **colony** of tardigrades could soon be living in another **habitat**.

Tardigrades can be hard to find because they're so small. But they could be living close to you, on moss, leaf litter or tree trunks, in sand or in other damp places.

Who's heading outside to say 'hello' to some of *Nature's Toughest Survivors?*

ABOUT TARDIGRADES

When scientist J.A.E. Goeze first saw tardigrades (pronounced TAR-dee-grades) under a microscope in 1773, he called them 'little water bears'. Four years later, another scientist called them 'Tardigrada', which means 'slow steppers'.

Tardigrades have been found in the most unexpected places – under damp rocks in deserts, in hot springs in Japan, on plants in forests and gardens, in rivers and oceans, and on the edges of a frozen lake in Antarctica.

They can survive extreme poisons that would kill any other organism – including cockroaches!

Water boils at 100° Celsius, but tardigrades have survived at the incredibly high temperature of 151° Celsius. The lowest temperature possible on Earth is −273.15. This is called absolute zero. Yet tardigrades have survived a temperature of −272° Celsius! They have even survived being shot out of a gun! Nobody knows exactly how long tardigrades can survive, but we do know that some tuns have lived for **decades**.

All tardigrades have a head and brain. Their barrel-shaped bodies have four sections. Each section has one pair of legs that end in sharp claws. Their last pair of legs faces backwards so they can anchor themselves when they are hunting or feeding. They have mouths and stomachs, and yes, they poo out waste!

Tardigrades don't have lungs. They get their oxygen from water, which they take in through their cuticles. Their bodies are filled with a fluid that carries oxygen and **nutrients** around their bodies, like our blood. Females can lay 1 to 30 eggs in their cuticles, which take 40 to 90 days to hatch. When baby tardigrades hatch, they look like little adults.

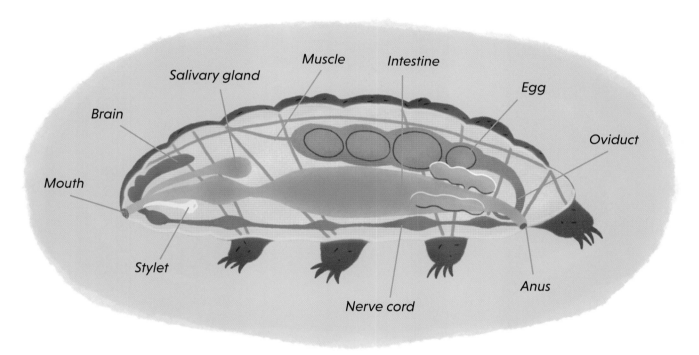

There are over 1300 tardigrade species. Some species have smooth cuticles, while others have tough, spiky cuticles. Most are less than 1 millimetre long.

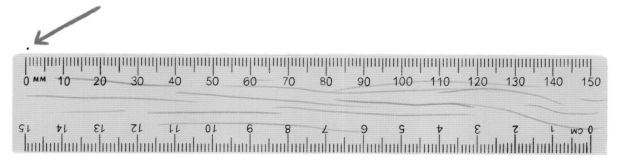

All over the world, scientists are studying tardigrades, hoping to learn from them. Tardigrades' ability to cope with changing temperatures, to recover from harmful events, and to put themselves into suspended animation could help researchers in many areas. These include making hardier **vaccines** and food crops, and improving space exploration. To survive such long journeys, human astronauts would have to put themselves in and out of suspended animation – like the 'tardinauts'.

GLOSSARY

Algae (AL-ghee): plant-like organisms that live in water and can photosynthesise (produce sugars from sunlight, water and carbon dioxide)

Bacteria (back-TEER-ee-ah): one of the oldest single-cell organisms on Earth; some bacteria help us make food such as cheese, or help us digest food, but other bacteria can cause disease

Cells: the 'building blocks' of plants, animals and other organisms; some, like bacteria, consist of only one cell, but tardigrade species may have up to 40,000 cells

Colony: a group of the same kind of plant or animal living together

Cosmic radiation: high-energy particles from beyond the solar system that can harm life

Cuticle (CUTE-i-kul): something that covers part or all of a living thing, for example, the bendy outer shell of tardigrades

Cyst: (in tardigrades) a shell of tough skin that contains an inactive tardigrade

Decade: 10 years

Fungi (FUNG-ghee): organisms such as mushrooms, moulds or yeasts that are neither plants nor animals; they live by breaking down dead plants or animals

Habitat: a place where a plant, animal or other organism naturally lives, such as the sea or a desert, which provides food, water and shelter

Microorganisms (MY-kroh-OR-guh-niz-ems): extremely small organisms that can't be seen without a microscope

Microscope: a piece of equipment that magnifies extremely small things

Nutrients: the parts of food that feed living things, providing them with energy to live and grow

Organism: an animal, plant or other life form

Oxygen: a gas needed by most living things to survive; also part of water, which all life requires

Species (SPEE-seas): a group of closely related organisms that can breed with each other

Stylet: a spear-like body part

Submersible (sub-MERS-i-bul): a small watercraft designed to work underwater, for research, repairs, exploration, etc.

Suspended animation: a state in which an animal is very close to death, but is able to bring itself back to life again

Vaccines: substances created to protect people from diseases

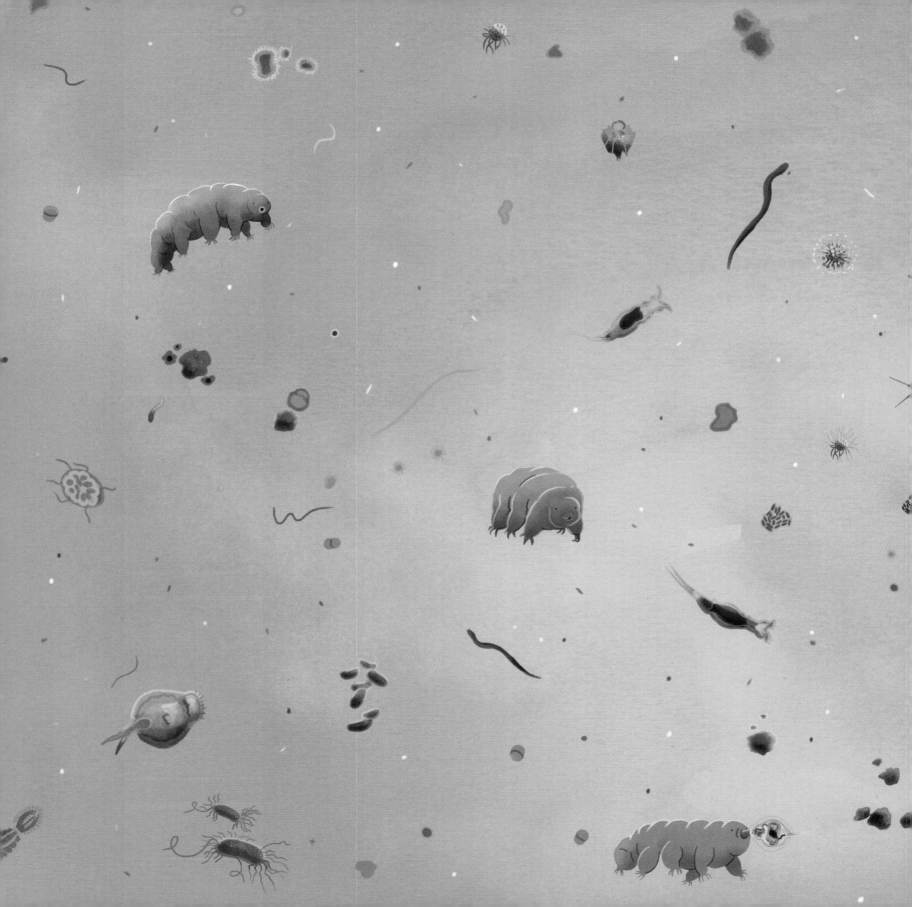